71+10 New
Science
Projects

Junior

Vikas Khatri

V&S PUBLISHERS

Published by:

V&S PUBLISHERS

F-2/16, Ansari Road, Daryaganj, New Delhi-110002
011-23240026, 011-23240027 • *Fax:* 011-23240028
Email: info@vspublishers.com • *Website:* www.vspublishers.com

Regional Offi ce : Hyderabad
5-1-707/1, Brij Bhawan (Beside Central Bank of India Lane)
Bank Street, Koti, Hyderabad - 500 095
040-24737290
E-mail: vspublishershyd@gmail.com

Branch Offi ce : Mumbai
Godown # 34 at The Model Co-Operative Housing, Society Ltd.,
"Sahakar Niwas", Ground Floor, Next to Sobo Central, Mumbai - 400 034
022-23510736
E-mail vspublishersmum@gmail.com

Follow us on: t f in

All books available at **www.vspublishers.com**

© **Copyright:** *V&S* PUBLISHERS
ISBN 978-93-505704-7-0
Edition 2014

Printed at : Unique Colour Carton, Mayapuri, Delhi

Publisher's Note

In line with a number of books and best-sellers in Science for children, particularly the school students, V&S Publishers have now come up with two new books on Science Projects and Activities namely **71+10 New Science Projects Junior** and **71+10 Science Activities**.

In the present book, **71+10 New Science Projects Junior**, the author has taken up the simple facts and principles of Science, such as: Air Pressure, Volume and Density, Gravitational Force of the Earth, Surface Area of Solids, Fun experiments with Matchsticks, Water, etc for children and projected them in a very simple and lucid language for the readers-particularly the school kids who can easily perform these experiments at home or school, of course with the help and guidance of their parents, elders or teachers.

Basically, the prime idea behind publishing such books are that they are educative as well as interesting and full of fun. Children of all ages, particularly from 6 to 13 can perform these fun-filled experiments and learn as well – the basic principles of Science easily and quickly. Moreover, Science, as we all know cannot be understood properly and learnt theoretically without practical demonstrations or experiments – in order to test and prove the different scientific laws and theories.

Therefore, this book will enable all its readers, especially the young ones to learn as well enjoy performing all the 81 Projects listed in the book, each explaining or proving some scientific theory or law – making it really worth reading!

Contents

10 Projects in Full Colour

1 Air, of Course

Things Required:

- String
- Wooden dowel
- Tape
- 2 paper bags
- Candle
- Matches
- THE HELP OF YOUR PARENTS/GUARDIAN/TEACHER

Directions

What is lighter than air? Do this experiment to find the answer.

1. Tie a piece of string to the centre of a wooden dowel and attach the free end of the string to a support, such as the centre of a doorway. Tape equal lengths of string to the bottom of 2 paper bags and tie these upside down to the ends of the dowel. Adjust the paper bags so that they balance perfectly on the dowel.

2. Hold one of the bags in its balanced position and ask one of your parents to light a candle and hold it under that bag for several seconds. When the candle is taken away, let go of the bag. The bag will rise.

Explanation:

The candle flame heats the air inside the bag. This hot air is surrounded outside by cooler air. *The cooler air is heavier than the hot air and lifts the bag of hot air upwards.* What is lighter than air? Air, of course!

2 Boomerang Returns to the Thrower

Things Required:

- Cardboard
- Pencil
- Scissors
- Book

Directions

1. Draw a V-shaped pattern like the one shown in the illustration on a piece of smooth, stiff cardboard. Cut it out, making sure the corners are rounded.

2. Hold a book in your left hand, with the binding pointing upward at an angle. Place the cardboard shape on the book so that one arm hangs off the side.
3. Hold a pencil along the side of the book. Quickly move the pencil forward and strike the cardboard so that it spins and flies off the book. In a few seconds, the cardboard will be back at your feet!

Explanation:

You have just made a *boomerang*. Because of its shape, the boomerang returns to the thrower, continuing to spin in the same path without turning over, Boomerangs, used by native Australians and usually carved from wood, can be used as weapons, or for hunting, or just for the fun of it.

3 Compressed Air

Things Required:

- Cork
- A glass soda bottle
- Petroleum jelly
- Water

Directions

1. Find a cork that will fit a glass soda bottle. Now, rub petroleum jelly around the sides of the cork.

2. Fill the bottle with water, about 1 inch from the top. Set the cork in the mouth of the bottle, but do not press it down. Instead, form a tight fist and slam the cork with one sudden blow. The cork will pop out. Now, press the cork in place slowly. It will stay there.

Explanation:

Scientists say that *air is elastic*. When the air is squeezed, it will press right back. This is what happened when you forced the cork down suddenly. The air left in the bottle was squeezed, so it pushed upward and sent the cork flying out. However, when you press the cork in slowly, the compressed air has time to leak through the seal between the cork and the glass.

4 Surface Area

Things Required:

✐ 2 sheets of paper

✐ A chair

Directions

This experiment takes only a few seconds to perform, but you may want to repeat it a few times and think about the scientific principle before reading the explanation at the bottom of the page.

1. Use two sheets of paper that are exactly alike. Crumple one piece of paper into a ball. Do not do anything to the other piece.

2. Stand on a chair and hold one piece of paper in each hand. Extend your arms as high as possible. Drop the crumpled piece of paper and the flat piece at the same time. Which paper falls faster? You know that they both weigh the same. Can you explain the difference in speed?

Explanation:

Even though both the pieces of paper weigh the same, they are shaped differently. The crumpled piece is more compact and is, therefore, able to push through the air better. The flat paper has more surface area and the air pushes against this and slows the paper down. Engineers who build aeroplanes and rockets know this scientific principle very well. They design their vehicles with a streamlined shape so that they can slice through the air instead of pushing against it.

5 Currents of Hot Air

Things Required:

- Aluminium plate
- Pencil
- Glue
- Empty spool
- Small block of wood
- Hot Plate

Directions

1. Cut out a spiral shape from the flat base of an aluminium pie plate. Use the blunt point of a pencil to make a dent in the exact centre, but do not poke it all the way through the aluminium.

2. Glue an empty spool to a scrap piece of wood. Insert the pencil, eraser side down, into the spool. The fit should be firm so the pencil stands upright without wobbling. If the pencil does wobble, wrap paper strips around it for padding.

3. Align the dent in the aluminium spiral with the pencil point and allow the spiral to hang freely. Twist or bend the metal as needed so that the edges separate from each other.

4. Finally, place your device over a safe heat source, such as a hot plate. The shimmering spiral will spin merrily.

Explanation:

You've just proven that hot air rises. The currents of hot air rise and push against the metal, and this continuous action causes the spiral to rotate around its pivot - the pencil point.

6 Clean Water

Things Required:

- 2.5 litre plastic jug
- Scissors
- Nail
- Hammer
- Pebbles, gravel and sand (coarse and fine)
- Glass jar
- Muddy water
- THE HELP OF YOUR PARENTS OR TEACHER

Directions

1. Cut off the bottom of a 2.5 litre plastic jug. Then, unscrew the cap and ask one of your parents to punch a few small holes in it with the tip of a nail and a hammer. Screw the cap back on and turn the jug upside down.

2. Fill the jug with equal layers of pebbles, gravel, coarse sand, and fine sand. The pebbles go in first, the gravel next, then the coarse sand, and, finally, the fine sand on top. Don't fill the jug completely; leave 5 cm free.

3. Hold the jug over a clear glass jar. If possible, prop up the jug so that it rests securely over the jar.

4. Now, pour some muddy water onto the sand. In a few minutes, clean water will trickle into the jar.

Explanation:

You have just performed *filtration*. Filtration is the removal of material, that is suspended in a liquid. The muddy water contained many impurities, and these were trapped - filtered - by the layers in your jug. The water itself, however, was free to pass through the layers and into the jar. Of course, you shouldn't drink this water because it is not really clean enough for drinking.

7 Excess Air

Things Required:

- A glass jar with lid
- Water
- 1 tablespoon salt

Directions

1. Fill a clear glass jar - such as a mayonnaise jar - with tap water. Set the jar in front of a bright window and watch the water at the top. Air bubbles will rise to the surface.

2. After the bubbling has stopped and the water becomes clear, add a tablespoon of salt to the jar. Screw the lid on the jar and turn the jar over once. Then return it to the upright position. Study the water again. More bubbles will rise. Where did they come from?

Explanation:

Water contains air, even though you can't see it. This air is usually dissolved in the water. You saw some of the excess air rising as bubbles in the first part of the experiment. When you added the salt, more air was driven from the water because the salt dissolves more easily in the water than the air, and it replaces the air. Fish in lakes and streams are able to take air directly from water by passing water through their gills.

8 Interesting Patterns

Things Required:

- Newspapers
- A half-litre can
- Water
- Enamel paint in several colours
- Small jam jars

Directions

This is a good project to do outdoors on a picnic table.

1. Spread some newspapers on the table on which you are working.

2. Fill a half-litre can about full with water, and then dribble different-coloured enamel paints into the can. You do not have to measure the paints exactly.

3. Hold a glass jar by the rim and dip it into the water. Remove the jar and set it upside down to dry. You will see beautiful swirls of colour that look like marble designs.

Explanation:

Enamel paints are made with oil and this causes them to float on top of the water. When you dip the jar into the water, the paint sticks to the glass and runs together to form interesting patterns. You might use your new jar to store rubber bands, seeds, paper clips, coins, or crayons.

9 A Drop of Water

Things Required:

- Stove or hot plate
- Frying pan
- Small glass
- Water
- THE HELP OF YOUR PARENTS, GUARDIAN OR TEACHER

Directions

1. Ask one of your parents to turn a stove burner or hot plate on high heat and to set a frying pan on the burner.
2. Fill a small glass, such as a juice glass, with water and set this down nearby.

3. After the pan has time to heat for a few minutes, with your parent at your side during this step, dip your fingers into the glass of water to moisten them. Shake the excess water from your fingers over the frying pan so that drops of water fall into the pan. You will see perfectly formed spheres of water bounce and parade around the pan.

4. Remind your parent to turn off the heat when you are finished.

Explanation:

As soon as a drop of water hits the hot surface of the frying pan, a little layer of steam forms underneath the drop. This steam acts as a cushion and raises the drop above the metal surface. The drop of water is held together as a sphere by the surface tension of the water, but eventually, the drop disappears, as all of its water is changed into steam by the heat.

10 Quality of Water

Things Required:

- Water faucet
- Spatula
- Spoon
- Small glass

Directions

1. Turn on the kitchen faucet and adjust the water to produce a smooth, continuous stream. Insert the flat portion of a spatula into the water's path. Hold the spatula horizontally and direct the water stream forward and downward until you produce a sheet of water.

2. You can make the water assume various shapes by slightly changing the angle and position of the spatula, or instead of the spatula, use a spoon to make different shapes from the water. If you hold the spoon rounded side up, you can produce a circular sheet of water!

3. Hold a small juice glass under the faucet. If you let the water strike the side of the glass at an angle, you can make a cone-shaped figure.

4. Search for other common household objects that might change the water's shape in unusual patterns. With a little practice, you can make many interesting shapes.

Explanation:

As you've seen, one of the many interesting qualities of water is its *surface tension*. When you inserted the objects into the stream of water, you spread the water over a wide area, but the water did not disperse. Instead, it held together in thin, clear sheets - This is due to surface tension!

11 Water Molecules

Things Required:

- Narrow drinking glass
- Water
- Cork
- Watering can (like the kind used for house plants)

Directions

1. Fill a narrow drinking glass with water to about 8-9 mm from the lip of the glass. Set a cork on the water surface and note that it drifts to the side. No matter how carefully you try to centre it, the cork will always move to the edge of the glass.

2. Now remove the cork. Using the watering can, or any other small pitcher with a spout, pour additional water slowly into the glass. Continue pouring until the water level is above the rim of the glass. Carefully place the cork on the surface once more. This time it will float in the centre.

Explanation:

In Step 1, where the water is 8-9 mm from the lip of the glass, the water clings to the walls of the glass. Because the water level is slightly higher at the walls than in the middle of the glass, the cork floats to this higher point. However, in Step 2, where the water is above the rim, the shape of the water is just the opposite - higher in the centre. So, again, the cork floats to the highest point, but this time it's the centre. Do you know why you were able to "pile up" the water above the normal level in Step 2? *If your answer was "surface tension," you're right!* The attraction of water molecules to each other allowed you to add water to the glass slightly above the normal level.

12 Surface Film

Things Required:

- String
- Water
- Cream pitcher, with a handle and spout
- Drinking glass

Directions

1. Cut a piece of string about 30 cm long and soak it in water for a few minutes.
2. Tie one end of the string to the handle of the cream pitcher, and then fill the pitcher with water.

3. Run the piece of string across the spout to the inside wall of the drinking glass. Press the string to the glass with your finger and pull the pitcher away until the string is tight. The pitcher should be several centimeters from the glass and slightly higher.

4. Now tilt the pitcher until the water pours out. The water will roll down the string and go into the glass.

Explanation:

The stream of water coming from the pitcher has a strong surface film around it. This film holds the water to the string, preventing it from dropping straight below. The string guides the path of the water and leads it into the glass. People who work in laboratories use this principle when they pour a solution from one container to another and do not spill a single drop. They usually place a glass rod across the spout of their pouring container and let the solution run along the rod into the other container.

13 Water is Heavier than Oil

Things Required:

- Small glass, such as a juice glass
- Vegetable oil
- Ice cube

Directions:

1. Fill a small glass with vegetable oil.
2. Place an ice cube in the glass, and you will see that the ice floats near the top. Observe your experiment for several minutes. As the ice melts, water droplets sink to the bottom. Do you know why this happens?

Explanation:

As you know that water and oil don't mix, and that since water is heavier, it will remain underneath the oil. So, since ice and water are made from the same matter, why did the ice float on top of the oil in this experiment? Well, even though ice and water are made of the same matter, each behaves in a different way. As water freezes, it expands and takes up more room. This makes it less dense and it flats in the oil. But once the ice has melted, the water is heavier than the oil and it falls to the bottom.

14 Candle Keeps Burning

Things Required:

- A glass soda bottle
- Water
- Small candle (like the kind put on birthday cakes)
- 2 or 3 straight pins
- Matches or Matchsticks
- THE HELP OF YOUR PARENTS, GUARDIAN OR TEACHER

Directions

1. Fill the soda bottle to the top with water.

2. Poke two straight pins into the bottom of a birthday candle and suspend the candle in the water. It should float upright. If the candle is tilted, you may need to add another pin to weight the base down a little more.

3. Now, ask one of your parents to light the candle. As you watch it burn down, stop and think: Will the flame die out when the wick burns down to the water level? Watch and see.

Explanation:

At the start of the experiment, the candle floats at the surface of the water. As the top burns away, the weight of the candle is decreased bit by bit. With less weight, the entire candle rises slightly, keeping the wick above the water level at all times. So, even though the candle grows shorter, the flame is never smothered by the water, and the candle keeps burning until its wick finally burns out.

15 A Rounded Surface

Things Required:

- A round balloon
- A string about 30 cm long
- A faucet

Directions:

1. Inflate a round balloon and tie the opening with a piece of string.
2. Turn on the faucet full force to produce a rushing stream of water.

3. Hold the end of the string and allow the balloon to hang freely. Slowly move your hand towards the faucet so that the balloon comes close to the water. As the balloon is drawn towards the water stream, pull your hand away slowly. The balloon will remain against the water jet and begin rotating.

Explanation:

The force of the water stream rushing into the sink creates an area of low pressure around it. Because the balloon is very light, it is pushed into this area by the surrounding air of higher pressure. The push of air on the rounded surface causes the balloon to spin.

Air Sucks

Things Required:

- A metal can
- Nail
- Hammer
- A round balloon
- Soap
- THE HELP OF YOUR PARENTS, GUARDIAN OR TEACHER

Directions:

1. Ask one of your parents to make a small hole near the bottom of a metal can by tapping the tip of a nail into the metal can with a hammer.

2. Inflate a round balloon until it is slightly larger than the can opening, and then tie the balloon shut.

3. Wet your hands and lather them with a piece of soap. Rub your soapy hands all over the surface of the balloon.

4. Place the metal can on its side on a table with the hole facing up. Hold the balloon next to the can opening and begin to suck the air from the tiny hole. The balloon will slip into the can. Now blow air into the hole, and you'll make the balloon leave.

Explanation:

By sucking air from the can, you decrease the air pressure inside. The air pressure outside the can is now greater, and this pushes against the balloon, forcing it into the can. Blowing into the can does just the opposite: The pressure builds up inside and forces the balloon out.

17 Burnoulli's Principle

Things Required:

- A bath spray hose
- Funnel
- Bathtub faucet
- Ping-Pong ball
- THE HELP OF YOUR PARENTS OR TEACHER

Directions:

1. Ask one of your parents to remove the spray head from a bath spray hose, and then insert the narrow end of the kitchen funnel into the hose opening.

2. Connect the other end of the hose to the bathtub faucet. Hold the funnel so that it points downward into the tub, and turn on the water.

3. Push a Ping-Pong ball into the funnel as far as you can. Now, take your hand away. The Ping-Pong ball will not be pushed out, but rather will stay securely in the funnel. If you turn the water on faster, the ball will only cling more firmly.

Explanation:

The stream of water rushing from the hose into the funnel produces an area of low pressure between the funnel and the Ping-Pong ball. The air pressure outside the funnel pushes upward on the ball and supports it against the downward thrust of the water. This experiment is a good example of *Bernoulli's principle:* The pressure in a flowing stream of liquid or gas is less than at its sides.

18 Water Rise

Things Required:

- A glass soda bottle
- Water
- Drinking glass

Directions

1. Fill a glass soda bottle till the rim of it with water. Hold a clear drinking glass upside down and place it over the mouth of the bottle.

2. Holding both the glass and the bottle together, turn them upside down at the same time. Some water may escape into the glass.

3. Raise the bottle 5 cm from the bottom of the glass and keep it in this position. You will see the water come from the bottle and go into the glass, but the water will stop as soon as it reaches the level of the bottle's mouth.

4. Repeat Step 3, raising the bottle 5 cm more. The water will never rise beyond the bottle. Can you explain why?

Explanation:

When you raise the bottle, air blows into it, pushing the water out of it. When the water level in the glass reaches the mouth of the bottle, the air outside the bottle presses on the water in the glass and prevents any more water from leaving the bottle.

19

Trapped Air

Things Required:

- A small, glass soda bottle
- Freezer
- Water
- A coin

Directions

1. Wash out an empty glass soda bottle and place it in a freezer.

2. After several hours, remove the bottle and moisten the top with water. Set a coin over the opening. The coin should make a seal at the mouth of the bottle.

3. Cup both hands around the sides of the bottle. Soon, the coin will jump up and down, tapping out a fascinating rhythm on the glass surface.

Explanation:

Cold air is trapped inside the bottle. As it begins to warm up, the air expands and forces the coin up. A little bit of air escapes and the coin falls back down. The process is repeated until the air inside the bottle reaches the same temperature as the air in the room.

20 Holding Together

Things Required:

- Newspaper
- Water
- A dinner plate
- Matches or Matchsticks
- A wide-mouthed jar
- THE HELP OF YOUR PARENTS OR TEACHER

Directions

1. Fold a newspaper page several times until it measures approximately 10 by 12.5 cm. Soak this piece in water, until it is completely wet, and then place it on a dinner plate.

2. Fold a smaller piece of dry newspaper, about 10 by 12.5 cm, into a narrow 1.25 cm-wide strip. Ask one of your parents to strike a match and light this strip. Then drop it into the wide-mouthed jar.

3. Ask your parent to quickly turn the jar upside down on top of the plate with the wet newspaper. Press firmly on the jar and continue holding this position until the flame has died out and the jar has cooled.

4. Now take the help of your parents or any gardian to hold the dinner plate in a grip and against the tabletop. Try to lift the jar. You can't - the jar remains fastened to the plate.

Explanation:

The burning strip of paper heats the air inside the jar. This hot air expands and some of it is forced from the jar. As the air remaining in the jar cools, it contracts and its pressure is reduced. The outside air pressing on top of the jar and underneath the plate is stronger than the inside air and holds the two objects together firmly.

21 Funnelling Device

Things Required:

- A funnel
- Glass bowl
- Water

Directions:

1. Set a funnel, small side up, on a counter top, and then set a large glass bowl next to it. Notice how high the tip of the funnel extends. Fill the glass bowl with water to a level just below the point, as shown in the figure.

2. Hold the funnel between your thumb and middle finger, keeping your index

finger over the small opening. Push the funnel into the water until it touches the bottom of the bowl. Raise your index finger slightly. You will feel a puff of air blown at your finger.

Explanation:

When you are holding the funnel with your finger over the tip, a quantity of air is inside the device. As you press the funnel to the bottom of the bowl, this air remains trapped inside the funnel, as your finger blocks it from the top and the water blocks it from below. When you remove your finger, the pressure of the water pushes against the air inside the funnel and forces it out through the small hole. If you used a glass funnel, you could see the water level rise inside the funnel as the puff of air hits your finger!

22 A Jet of Air

Things Required:

- A paper straw
- Scissors
- Drinking glass
- Water

Directions

1. Slit a paper straw across, from one end, making sure you do not slit it all the way through. Bend the straw back like a hinge and insert the short end into a drinking glass.

2. Fill the glass with water until the level almost reaches the hinged part of the straw. Now blow hard through the long section of the straw. Water will spray from the glass.

Explanation:

When you blow through the straw, the swiftly moving jet of air *reduces the pressure* above the hinged opening. The air pressure over the water is now greater, and this pushes the water up the short section. When this water hits the stream of air, it is carried away as drops of water. This is the same principle used in many squeeze-type spray bottles. Instead of blowing through a straw, however, you force air through the mechanism with a hand pump.

23 An Attached Ice Cube

Things Required:

- A glass
- Water
- An ice cube
- A string
- Some salt

Directions:

1. Fill a glass with water and place an ice cube on the surface.

2. Tie a loop, about 2.5 cm in diameter, in a piece of string several centimeters long. Set the loop on top of the ice cube.

3. Sprinkle some salt over the top of the cube where the loop sits. Wait a few minutes. Gently pull the string up. The ice cube is lifted above the water.

Explanation:

The salt caused the ice cube to melt around the string. Then the water refroze, freezing the string to the ice cube and allowing you to pull it up with the ice cube attached. Salt is used on many roads and sidewalks in the wintertime on ice and snow because it lowers the melting point of water.

24 Heavy Clouds

Things Required:

- Saucepan
- Water
- Frying pan
- Ice Cubes
- THE HELP OF YOUR PARENTS/TEACHER

Directions

1. Fill a saucepan about full with water, and ask one of your parents to boil it over high heat.

2. Fill a frying pan with ice cubes, and then give it to one of your parents to hold several centimeters above the steam escaping from the saucepan. Remind your parent to hold the frying pan by its handle so that his or her hands do not come in contact with the steam. Steam is extremely hot and can burn! In a few minutes, you will see raindrops fall from the bottom of the frying pan into the boiling water.

Explanation:

You have just produced *rain* the same way that nature makes it. The boiling water caused water vapour (steam) to rise from the saucepan. As the steam hit the cold surface of the frying pan, it collected as moisture on the underside. Soon, the moisture became too heavy and fell as drops of water. In nature, a similar cycle takes place. Oceans, lakes and streams lose water through *evaporation*. The water vapour rises into the sky. Here, it is colder, so the water collects into clouds. When the clouds become too heavy with water, drops of rain fall to the earth.

25 Stratus and Cumulus Clouds

Things Required:

- Fish tank
- Water
- Blue food colouring
- 1 cup vegetable oil
- Stirrer

Directions:

1. Fill a fish tank or other large, clear container about half full with water, and add some food colouring until the water turns to a deep shade of blue. Then pour the oil on top.

2. Slowly stir the water. The oil will stay mostly flat near the top. Next, stir the water very rapidly. The oil will roll over and form a fluffy appearance. Do these shapes remind you of anything?

Explanation:

You have just made a *model of the formations that clouds make in the sky*! When the air is calm, the clouds assume a level shape, like the oil that you first stirred. These kinds of clouds are called *stratus*. When the air is moving fast, however, the clouds roll over themselves, just as the oil did. Clouds like these are called *cumulus*.

26 Bolt of Lightning

Things Required:

✎ A thunder and lightning storm

Directions

Can you tell how far away a storm is? Here is a simple way to find out.

1. The next time a big thunderstorm occurs, watch for the lightning. As soon as you see the *flash in the sky*, start counting, "thundercracker 1, thundercracker 2, thundercracker 3," and so on. (The time it takes you to say "thundercracker," followed by the number, equals about a second.) Stop counting when you hear the clap of thunder.

2. Now divide the number of seconds you have counted by 5. The result will be the distance of the storm centre. For example, suppose you had counted to "thundercracker 10" when you heard the big boom: 10 5 = 2. The storm is about 3 km away.

3. You can repeat the procedure on the next bolt of lightning. If the storm is closer this time, you know that it is travelling toward you. Better get inside!

Explanation:

Light travels at a speed of about 2,97,600 km per second, so you see a bolt of lightning almost instantly when it occurs. Sound, however, travels much more slowly - at a speed of only 320 meter per second. When you see a bolt of lightning, you know that the sound has just started to travel. By determining how long it takes to reach your ears, you can figure out how far away it was.

27 Tube Strength

Things Required:

- A sheet of typing paper
- Rubber band
- Book

Directions

1. Roll a single sheet of typing paper into a tube and slip a rubber band around it.

2. Stand the tube one end on a flat surface. Carefully place a book on top of the tube and you will see that the paper supports the weight of the book.

Explanation:

A tube is a shape that has much more strength than a flat object. This allows you to place the book on top of the paper without crushing it. Pillars are also of tube shape, and they are used in some buildings to hold up their great weight.

28 Scattered Force

Things Required:

- Saw
- Board
- A small glass bottle with a protruding lip
- Metal trash can
- Water
- Cork
- Hammer
- THE HELP OF YOUR PARENTS, GUARDIAN OR TEACHER

Directions

1. Ask one of your parents to use the saw to cut a notch in a long, flat board so that the bottle can be suspended from it by resting the glass lip on the wood.

2. Lay the board across a metal trash can. Fill the bottle to the top with water and insert the cork. Make sure there is no trapped air (no air bubbles) inside. Set the bottle in the notch.

3. Now, ask one of your parents to tap the cork with a hammer, tapping a little bit harder with each strike of the hammer. With very little force, the bottle will shatter into the trash can. Of course, no one should try to pick up the broken pieces - leave them in the trash can.

Explanation:

When the cork is hit with the hammer, a force is created that is transmitted into the water. Since the water is confined to a single area, the force is scattered throughout the substance in all directions. The walls of the glass bottle cannot withstand this great pressure, and they break.

29 Tightly Packed Grains

Things Required:

- An empty mayonnaise jar
- Uncooked rice
- Knife with blunt, wide blade, such as a cake knife

Directions

1. Fill the empty mayonnaise jar with uncooked rice and pack it down firmly. Add more rice until it's even with the top of the jar.
2. Poke the blunt knife into the rice several times to a depth of about 5 centimeters. Then jab the knife in firmly, about 15 cm deep.
3. Now slowly pull the knife upward. You will lift the jar of rice.

Explanation:

The rice grains, which fill the jar, have many air spaces between them. As you poke the knife into these grains, they become tightly packed. When you finally jab the knife deeply, the rice is pushed against the blade holding it in place. This gripping force enables you to lift the entire jar as you raise the knife.

30 Drifting Cardboard

Things Required:

- An empty food can, such as a soup can
- A can opener
- Scissors
- Cardboard
- Pail
- Water
- Drinking glass
- THE HELP OF YOUR PARENTS, GUARDIAN OR TEACHER

Directions:

1. Ask one of your parents to cut away the bottom of an empty food can with a can opener so that both the ends are open.

2. Cut a piece of cardboard that is a little bit larger than the bottom of the can.

3. Fill a pail with water. Hold the cardboard beneath the bottom of the can and push the can straight into the water. When the outside water level comes near the top of the can, take away your hand from the cardboard. The inside of the can will remain dry as the cardboard clings to the can.

4. Now pour water slowly from a drinking glass into the can. When the water level inside the can is the same as the water level outside, the cardboard will break away.

Explanation:

The empty can and cardboard act as if they were one solid unit in the water – the force of the water presses upward on the cardboard, keeping it pressed to the bottom rim of the can. However, when you add water to the inside of the can, you are creating an opposite force – a downward force, which balances out the water pressure in the pail, and the cardboard drifts away.

31 Molecules in Motion

Things Required:

- A small bar of soap, not the kind that floats
- A glass jar with lid
- Glue
- Paper
- Water
- Pencil

Directions:

1. The tiny bars of soap from hotels or aeroplanes are excellent for this experiment, but if you don't have any, a slice or a piece of regular soap broken into several chunks will cover the bottom of the jar.

2. Glue a strip of paper up the side of the jar. Then drop the soap into the jar, and fill completely with water.

3. Screw the lid onto the jar and set the experiment in a quiet place where it will not be disturbed. Check the experiment every week for several weeks. You will see two layers in the jar.

The soap dissolves to form a heavy solution underneath the water. Mark the soap position on the paper each week. This layer slowly creeps upward. Do you know why?

Explanation:

At first, the soap dissolves in the water surrounding it. This is why you see the layer of soap solution at the bottom. However, the molecules of a substance are always in motion, even though the substance may appear to be sitting quietly. The soap and water molecules are in constant motion, always interacting. Eventually, the soap solution distributes itself throughout the entire jar of water. The scientific name for this process is *diffusion*.

32 Energy Charge

Things Required:

- Tape
- String
- Coin
- Desk

Directions

1. Tape a piece of string to one side of a coin. Tie the free end of the string to the handle on an opened desk drawer.

2. Keeping the string straight, pull the coin back until it touches the front of the desk. Then release it, letting it swing away. Observe its motion. You will see that the coin travels over a shorter distance with each swing it takes.

Explanation:

For any action, the energy out always equals to the energy going in. So, when you start the coin at the desk, it never swings higher because this would require more energy. But you would expect that the coin should travel the same distance in each swing. Why does it swing lower? Energy is never lost; however, some energy changes to a different form. The act of the coin rubbing against the air is called *friction*, and friction changes the *energy of the swinging coin into heat*. The surrounding air actually becomes warmer, but this change is so slight that you don't notice a temperature increase.

33 Rise in Temperature

Things Required:

✎ A heavy-duty rubber band, at least 6 cm wide

Directions:

1. Hold the rubber band between your two hands and stretch it tightly. Gently place it against your cheek. The rubber band feels warm.

2. Now release the tension in the rubber band, and once again hold it to your face. This time it is cool. Repeat the stretching and loosening process several times. Do you know what causes the gain and loss of heat?

Explanation:

If you could not come up with an explanation for this experiment, don't worry. Scientists have several theories, but no one is sure which is absolutely correct! One idea is that when the rubber is stretched, the molecules bump into each other more frequently and this may raise the temperature.

34 Candle Flame

Things Required:

- A candle
- An aluminium pie tin
- A coin
- An index card
- Matches or Matchsticks
- THE HELP OF YOUR PARENTS, GUARDIAN OR TEACHER

Directions:

1. Set the candle on the aluminium pie tin and place the tin in the sink.
2. Rest the coin in the centre of the index card.

3. Ask one of your parents to light the candle and move the card over the tip of the flame, keeping the card in continuous motion, but the coin in the same position. When the paper begins to turn brown, ask your parent to stop.

4. Let the coin slide off the card into the aluminium dish. The coin will be hot, so do not pick it up until it has cooled. You will see a pattern on the card in the spot where the coin was resting.

Explanation:

The change in colour of the paper is due to *the heat from the candle flame*. A slight, charring takes place. The area where the coin is resting, however, remains unburnt because the metal conducts heat away from that space.

35 Yarn Quality

Things Required:

- A nylon rope, or nylon clothesline, 2 to 3 metres long
- Dacron, polyester, or dacron/polyester rope, 2 to 3 meter long
- Water
- 2 big rocks
- THE HELP OF YOUR MOTHER OR FATHER OR TEACHER

Directions:

1. Wet both pieces of rope with water from a garden hose or in a sink.

2. Ask one of your parents to tie each rope to a tall, sturdy object, such as the cross bar of a swing or a high, solid tree branch.

3. Tie a large rock to each of the free ends. Now tug on the ropes by applying downward pressure to the rocks. What do you notice about the properties of the two different kinds of rope? The nylon line increases in length, while the dacron line remains the same.

Explanation:

Sailors and other boaters know about the two different kinds of rope that you tested. A *Nylon line* will stretch a little while it is under tension, so sailors use it to tie their boats to a dock. However, the lines used to fasten the sails to the mast must be as tight as possible so the sails don't flap needlessly in the wind, and a stretchy rope would not be a good idea. In this case, *Dacron line* is used.

36 An Amazing Match

Things Required:

- A wooden match
- A large safety pin

Directions:

1. Break the striking head off the wooden match and discard it. Open the safety pin and push its point through the centre of the match. Close the pin. Move the match around the metal several times so that it rotates easily.

2. Now rotate the match until it is pressed against the other edge (on top) of the pin. Push firmly against the lower tip of the match, and quickly slide your finger off the edge in a snapping motion. The match seems to move through the solid pin to the other side.

Explanation:

Of course, the wooden match cannot pass through another solid material. When you pressed the tip of the match, this caused the stick to snap against the safety pin and bounce around in a full circle until it came to rest on the opposite side of the metal. This happened so fast that it appeared as if the match passed through the metal. Practise this trick several times and show it to your friends. They'll be amazed!

37 Uniform Density

Things Required:

- A raw egg
- A hard-boiled egg

Directions:

Can you tell the difference between a raw egg and a hard-boiled egg? They look and feel the same, but here is an easy trick.

1. Spin a raw egg on a hard surface, such as a counter or tabletop. (Don't let it fall off the table or you'll have quite a mess to clean up!) The egg will slow down very soon and move in a floppy, random fashion.

2. Now spin a hard-boiled egg. The egg acts quite differently this time. It will spin easily and may stand up one end. You will also notice that it spins for a much longer period of time.

RAW

HARD BOILED

Explanation:

The hard-boiled egg is of *nearly uniform density* throughout its interior. The raw egg, on the other hand, has a loose, runny composition, and the shifting contents slow down the motion of the egg.

38 Concentrated Stress

Things Required:

✐ Cellophane packaging material

Directions:

1. Save the plastic wrapper from a packaged food item, such as a bag of pretzels.
2. Try to make a rip in the cellophane by poking your fingers into it, then pulling it apart. You will find that it is very hard to start a tear, but it is easy to keep it going after it has started.

Explanation:

When you pull against the opposite ends of the plastic, your *force is spread over a large area*. The plastic may stretch, so starting a tear is very hard. Once a tear has begun, however, most of the stress is concentrated at this single point, and you do not have to pull very hard to continue the tear.

Things Required:

- A small paper bag (lunch bag)
- Pail
- Water
- A string

Directions:

1. Set the small paper bag into a pail filled with water.

2. Allow the bag to fill with water, and then tie it closed with a piece of string. The bag will drift in the water without any harm.

3. Now raise the bag from the water by pulling on the string. The paper will burst open immediately. Can you explain why?

Explanation:

While in the pail, *the paper bag is surrounded on all sides by a uniform medium*. That is, all forces are balanced and there is no strain in any particular spot. However, as the bag is raised, the air is less dense than the water inside the bag. Gravity pulls downwards on the water, and the bag is not able to withstand this force.

40 Escaping Sand

Things Required:

- A coffee can with plastic lid
- Hammer
- A small nail
- A string
- Sand
- A large sheet of paper
- THE HELP OF YOUR PARENTS, GUARDIAN OR TEACHER

Directions:

1. Ask one of your parents to punch a hole in the bottom of the coffee can with

a hammer and a small nail. Then make 3 equally spaced holes around the top edge of the can.

2. Tie a piece of string, approximately 15 cm long, through each hole, and knot the ends together at the top. Tie a long piece of string to this knot and hang the can from a low tree branch. The bottom of the can should be 2.5 or 5 cm above the ground.

3. Cover the bottom of the can with the plastic lid and fill the can with clean, dry sand.

4. Spread a large sheet of paper on the ground. Remove the plastic lid and give the can a push. You will see a unique sand pattern form in front of you.

Explanation:

You have just made a *pendulum*. Its action is traced by the escaping sand. The two parts of the swinging motion - vertical and horizontal - are combined into the pattern that you see on the paper.

41 Gas Bubbles

Things Required:

- Knife
- Cooked spaghetti
- A fish bowl
- 1 cup vinegar
- 1 cup water
- Red and blue food colouring
- 2 tablespoons of baking soda or sodium bicarbonate ($NaHCO_3$)
- THE HELP OF YOUR PARENTS OR TEACHER

Directions:

1. The next time you have spaghetti for dinner, ask one of your parents to cut a few strands of the cooked spaghetti into 2.5-6.5 cm pieces.

2. In a fish bowl or other large container, mix the cup of vinegar and the cup of water. Then add a few drops of red and blue food colouring. Slowly add the 2 tablespoons of baking soda.

3. Now drop the pieces of cooked spaghetti into the bowl. Purple worms will waggle back and forth, but some will rise to the top and fall to the bottom of the container several times. Can you explain why?

Explanation:

The vinegar and baking soda form gas bubbles, which collect on the spaghetti. Because the gas bubbles make the spaghetti lighter, the pieces rise and drift in the solution. The gas bubbles of those pieces that rise to the top of the container break open at the surface, causing the pieces to fall to the bottom, where more gas bubbles collect on the spaghetti, and the process is repeated.

42 A Complete Path

Things Required:

- An insulated wire (the kind with the plastic coating)
- A flashlight bulb
- D-size battery
- Masking tape
- THE HELP OF YOUR PARENTS, GUARDIAN OR TEACHER

Directions:

An electrical conductor is a material through which electricity flows. You can build a device that will test different materials to find out whether or not they are electrical conductors.

1. Ask one of your parents to do the following: Cut two 25 cm lengths of insulated wire. Strip about 8 cm of coating from one end of one piece of wire and wrap this tightly around the base of a flashlight bulb. Strip about 15 mm of coating from the remaining ends.

2. Place the flashlight bulb at the pointed tip of a D-size battery. Tape one end of the loose wire at the flat end of the battery. Tape the remaining wire in place on the battery, as shown.

3. Now tape the two free wire ends 15 mm apart, placing the tape on the coated portion of the wire, leaving the bare ends free.

4. To operate your tester, press the bulb firmly against the battery. Touch the two bare wires to the object being tested. If the material is a conductor, the bulb will flash. Try testing a pair of scissors, this book, and your bicycle!

Explanation:

Electricity is generated in the battery and this flows into the wires. When you touch the ends to a conductor, a *circuit* is completed. This means that the electricity can now flow through a complete path. So the electricity flows through the wires, the object, and the bulb, and the bulb lights up.

43 Mild Current

Things Required:

- A copper nail
- A zinc nail
- Steel wool
- Lemon

Directions

1. Scrub a copper nail and a zinc nail with a piece of steel wool until they are clean and shiny. Rinse the nails under the faucet.
2. Now poke the pointed ends of the nails into the centre of a fresh lemon. Space the nails about 2.5 cm apart and leave half of each nail protruding.
3. Stick out your tongue and touch it across the tops of the nails. You will feel a tingle.

Explanation:

You have just made a *chemical battery* and the tingle on your tongue was *electricity*. The lemon contains acid and water, which reacts with the metals, copper and zinc to produce a slight current that passes over your tongue.

44 An Opposite Charge

Things Required:

- A wool sock
- Plastic comb
- Tape
- Lightweight thread, about 30 cm long

Directions:

1. Rub a wool sock rapidly back and forth over a plastic comb.
2. Tape the end of the piece of thread to the top of a table.
3. Touch the comb to the free end of the thread and raise the comb. The thread will stand on end, sticking straight up into the air.

Explanation:

When you rub the wool sock over the comb, you give the *comb an electrical charge*. The charge on the comb attracts the thread because the thread carries an opposite charge. The two differently charged objects cling together, and the thread is carried upward with the comb.

45 A Model of Space

Things Required:

- A round balloon
- A wide-tip felt marker

Directions:

1. Inflate a round balloon partially full with air. Pinch the neck closed with your thumb and forefinger, but do not tie it closed.

2. With a wide-tip felt marker, make several specks all over the surface of the rubber and let them dry.

3. Now blow more air into the balloon, take it away from your mouth, and note the position of the specks. Continue inflating and observing the balloon. The specks grow farther and farther apart. Do you know what this model represents?

Explanation:

The *balloon is really a model of space, and each of the spots is a galaxy of stars*. Our Sun and the planet, Earth are part of the Milky Way galaxy. Scientists believe that the universe is expanding in just the same way as you saw on the balloon. The galaxies are drifting apart, leaving greater distances between them.

46 A Cloud Chamber

Things Required:

- Scissors
- Sponge
- Mayonnaise jar with lid
- Glue
- Tin snips
- Carbon paper
- Rubbing alcohol
- Dry ice
- Slide projector
- THE HELP OF YOUR PARENTS, GUARDIAN OR TEACHER

Directions

1. Cut a piece of sponge so that if fits into the bottom of a mayonnaise jar. Glue the sponge in place.

2. Ask one of your parents to cut out a 2.5 cm piece of metal from the side of the jar's lid with a pair of tin snips. Do not handle the edges because they will be sharp.

3. Next, cut a piece of carbon paper into a circle that will fit into the metal lid. Place the paper in the lid with the carbon side up.

4. Pour some rubbing alcohol into the jar and let the sponge soak up as much of this liquid as it can. Then turn the jar upside down, and let it drain completely. Now, crew on the lid. Ask one of your parents to set the device, still upside down, on a small piece of dry ice.

5. Shine a projector light into the cutaway section of the lid. In about 10 minutes, you will see tiny trails skirting across the black background of the carbon.

Explanation:

You have just constructed a *diffusion cloud chamber*. This instrument makes the paths of nuclear particles visible. Nuclear particles may come from outer space, the earth's natural radioactivity, or from man-made sources, such as power plants. As the alcohol in the sponge vaporises, it slowly sinks to the lid where it is cooled by the dry ice underneath. The motion of nuclear particles through this cool vapour creates trails of fog, which you are able to see because of the bright light.

47 The Earth's Axis

Things Required:

- 2 straight pins
- A 1 or 2-rupee coin

Directions

1. Place the points of 2 pins at directly opposite edges of a 1 or 2-rupee coin. Keep the pins steady and straight, and gently lift. You may need to try several times before you can pick up the quarter without it slipping. If someone is nearby, perhaps you could ask him or her to hold the coin for you while you position the pins.

2. Once the coin is securely held between the pins, blow at the top half. The coin will spin rapidly.

Explanation:

A spinning object revolves around a line, and this line is called the *object's axis*. You have formed an axis through the coin. This is the line extending straight between the pins. *The earth spins around an axis also, but, of course, there are no pins supporting it. The line is imaginary.*

48 North and South

Things Required:

- Needle
- Magnet
- Scissors
- An index card
- Jar
- Thread
- Pencil

Directions

1. Magnetise a needle by stroking it several times with a magnet.
2. Cut a small strip from an index card so that the strip will fit inside the jar. Push the needle into the card.
3. Tie one end of a piece of thread to the centre of the strip, the other end, to

a pencil. Suspend the strip inside the jar by resting the pencil across the opening.

4. The needle should rest horizontally - you can balance it by sliding it back or forth in the card strip.

5. Let the device come to rest. You have made a *compass*, and the *needle* will point *North and South*.

Explanation:

The entire *earth has a magnetic field* surrounding it. The needle is a miniature magnet and it is attracted by the earth's magnetic forces. Since the needle is free to rotate, it aligns itself in a *North and South direction*.

49 Suspended Liquid

Things Required:

- Flavoured gelatin dessert
- Small dish
- Eyedropper
- Water
- Fork

Directions

1. Pour a box of flavoured gelatin into a small dish. The powder should be at least 2.5 cm deep.

2. Using an eyedropper, squeeze a drop of water onto the surface and let it soak in. Then continue squeezing single drops of water onto the same spot until 6 drops have been deposited. Allow each drop to soak in before adding the next.

3. Now dip a fork under this area and gently lift upward. A chewy little gumdrop will come to the surface.

Explanation:

Gelatin dessert is made of *sugar, flavouring and protein*. As you add drops of water to this dry powder, the mixture swells and holds the water in place, keeping the liquid suspended within the surrounding solid material with its protein fibres.

50 True Solution

Things Required:

- Red and blue food colouring
- ½ cup water
- Bowl
- 1 heaping cup cornstarch
- 2 marbles

Directions

1. Add a drop of red and a drop of blue food colouring to half cup of water. Pour this into a bowl. Add the cornstarch and mix well.

2. Pick up a handful of this goo (a thick, sticky substance) and roll it quickly

between your hands, forming a ball. The solution feels dry. Stop the rolling action, and the substance loses its form, oozing between your fingers.

This experiment is a good stunt to show your friends – they'll be amazed!

Press two marbles into your purple concoction, and it will look like you've made a new kind of creature!

Explanation:

Corn-starch does not form a true solution with water. Instead, solid particles are held up by the water, creating a mixture called a *suspension*. When you roll the mixture in your hands, you keep the suspension together by squeezing it on all sides. But as soon as you remove this support, the fluid and its particles are able to flow freely, drifting apart. If you let some of the goo sit for several minutes in a clear glass, you will see the cornstarch and water separate into two layers.

51 Garden without Plants

Things Required:

- A cup of water
- 1 cup laundry blue
- A cup of salt
- 1 tablespoon Ammonia (NH_3)
- Jar
- Spoon
- Charcoal briquettes
- Bowl
- Food colouring in various colours
- An old pie tin

Directions

1. Place the water, laundry blue, salt and ammonia in a jar, and stir thoroughly with a spoon.

2. Set a single layer of charcoal briquettes in a bowl. Then pour the solution on top. The solution should not cover the charcoal completely.

3. Put several drops of various shades of food colouring on the charcoal, leaving some areas plain.

4. Set the bowl in an old pie tin and place the entire experiment in a quiet place. The next day you will see gorgeous crystal formations covering the charcoal and sides of the bowl.

Explanation:

There are many *small spaces inside the charcoal briquettes*, and the solution was drawn into these areas. As the water evaporated, the salt remained there, forming crystals. The crystals have similar spaces themselves, and the solution continued to be sucked up and evaporated. Thus the delicate formations continued to grow by attaching more salt to the ends of the existing crystals.

5² Air in the Egg

Things Required:

- A fresh egg
- A bowl
- Hot tap water

Directions

1. Place the fresh egg in the bowl, then fill the bowl with hot tap water.
2. Set the bowl on a table or counter top and watch closely as the experiment sits quietly for several moments. You will see a tiny stream of bubbles rising from the egg.

Explanation:

Did you know that an egg contains air? The air inside the egg expands as it is heated by the hot water, and escapes into the water as bubbles. You might wonder how the air leaves the egg since the shell is not broken. Well, there are tiny openings called *pores in the shell* - about 7,000 of them! The pores are big enough to allow gases and moisture to pass through, but small enough to prevent harmful bacteria from getting into the egg.

53 Nutty Fat

Things Required:

- A brown paper bag
- Scissors
- Window
- Peanut

Directions

1. Cut open a brown paper bag, such as a shopping bag. Lay the flat piece against a window.

2. Rub a peanut, without its shell, firmly on the bag. Move the nut back and forth several times over the same area. Soon you will see light coming through the rubbed space.

Explanation:

Did you know that peanuts contain a large amount of fat? When you rub the peanut on the bag, the bag absorbs some of this fat, which soon spreads into the paper and fills in the spaces between the fibres. This allows light to pass easily through the paper.

54 A Sprouting Pit

Things Required:

- A ripe Avocado pit
- 3 wooden toothpicks
- A small juice glass
- Water

Directions:

1. Wash off the ripe Avocado pit and peel away the dark brown coating.

2. Insert three wooden toothpicks, equally spaced, around the middle of the pit. Then set the pit in a small juice glass so that the toothpicks are resting on the glass rim.

3. Fill the glass with water to a level just beneath the toothpicks. The bottom of the pit should be covered with water.

4. Set your experiment aside for several days, but maintain the water level by adding more water when necessary. This last step will vary in time, depending on your pit. In a few days to a few weeks, you will see the pit split in half. A root will come from the bottom, and a sprout will grow from the crack.

Explanation:

An avocado pit is just like any other seed, except that it is larger and harder. Everything needed to make a new avocado plant is contained in the pit. After the root and sprout emerge, the two halves of the pit supply food for the growing plant until it grows leaves and starts to make its own food from sunlight. You can plant your new avocado in a pot with soil to make a nice house plant.

55 Weight Loss

Things Required:

- A large potato
- A small, sensitive scale, such as a postal or dieter's food scale

Directions:

1. Wash and dry the potato, then place it on the scale. Note the weight and write it down.
2. Now set the potato in a dry place and leave it there for about three weeks.
3. After three weeks, weigh the potato again. Even though the potato may look the same size to you, it will weigh less than when you first weighed it. Has the potato gone on a diet?

Explanation:

Water makes up a large part of most animal and plant matter. When the potato was exposed to the dry air, some water evaporated and this resulted in a weight loss. Did you know that most of your own body is made up of water?

56

Plant Power

Things Required:

- 10 cm-high flowerpot
- Soil
- 10 corn seeds
- Water
- Glass plate

Directions

1. Fill the flowerpot with soil almost to the top. Poke about 10 corn seeds into the soil, then sprinkle more soil on top.

2. Water the seeds thoroughly and place the pot in a warm place. As the soil dries out, give it water.

3. After the seeds have sprouted in a few days, cover the pot with a piece of

glass slightly larger than the top. Allow the corn to continue growing with the glass in place. Soon you will see the glass plate lifted from the rim of the pot.

Explanation:

We usually don't think of plants as having muscles as humans do, because *plants remain still, while we are very active*. But over a period of time, a growing plant can exert a tremendous force. Have you ever seen a cement sidewalk pushed up by the growing roots of a big tree?

57 Skeletons under the Sheet!

Things Required:

- Leaves
- Newspaper
- Books
- Scrap paper
- Hammer

Directions:

1. Collect several kinds of leaves and place them between the folds of a newspaper. Then put some heavy books on top. Keep the leaves here for a few days, until they become dry and brittle.

2. After the leaves have dried out, remove one from the newspaper and place it between the sheets of scrap paper. Then use a hammer to gently pound the entire leaf area.

3. Remove the top sheet of paper and lift the leaf by its stem. You are holding a leaf skeleton.

Explanation:

After being dried out and pounded, most of the plant cells have crumbled and fallen away, leaving only the veins. While the leaf was alive and growing, these veins transported food and water within the plant, and provided the leaf with a solid framework. Make *leaf skeletons* from the other leaves you collected and notice how their shapes differ.

58

A Growing Mold

Things Required:

- A slice of bread
- 2 paper plates
- A spray bottle
- Water
- Magnifying glass

Directions

1. Place a slice of bread on a paper plate. Homemade bread or one of the natural breads without preservatives works best, but ordinary bread will also do the job.

2. Fill a spray bottle with water and gently mist the top of the bread until it is moist. Do not soak the entire slice.

3. Let the bread sit in the open air for a few minutes. Then set another paper plate, upside down, over the first. This will form a raised cover for the bread. Leave the entire experiment in a warm, dark place for several days.
4. You will soon see a light gray fuzzy material covering the bread. If you check the bread each day now, the growth will become heavier and turn dark.

Explanation:

You probably knew that the material growing on the bread is called mold – a unique kind of plant. *Mold is not green and does not make its own food in sunlight.* In this experiment, the mold drew nourishment from the bread. Even though mold can spoil food, sometimes it is helpful, and people grow it on purpose like you did. Many cheeses and certain drugs, such as *penicillin*, are made from molds.

Use a magnifying glass and examine your mold carefully. Do you see small dark spheres on the ends of stalks? These structures help to grow new molds. Tear the bread apart and look at the torn edge. There are several white threads growing into the bread. These are similar to the roots of a plant.

59 Don't Play with Mother Nature

Things Required:

- A winter day
- A Forsythia shrub
- Pruning shears
- A large vase
- Water

Directions

Here is an experiment that will help you brighten up the grey days of winter.

1. If you don't know what a forsythia shrub looks like, ask an adult to point one out. You might have one of these shrubs right in your own backyard! With pruning shears, cut a few branches from the bush and bring them indoors.

2. Fill a large vase with water. Place the branches upright in the water and set the vase near a sunny window. In a few days, you will see bright yellow flowers along the length of the branches.

Explanation:

You have just played a trick on Mother Nature. In the fall months, forsythia branches form buds that contain flowers. These buds lie dormant all winter long, and with the spring's warmth and moisture, the buds grow and develop into beautiful flowers. You speeded up this process by bringing the branches in from the cold. *The water in the vase and the warmth and light of the room fooled the Forsythia branches into thinking, it was spring.*

60 Paleontology

Things Required:

- 2 plastic margarine tubs
- Plaster of Paris
- Old spoon
- Water
- Petroleum jelly
- Small seashells

Directions:

1. Fill a plastic margarine tub about full with plaster of Paris. While stirring with an old spoon, slowly add water until the mixture becomes creamy.

2. Spread a thin coat of petroleum jelly onto the outside of several seashells. Press each shell into the plaster, but do not allow the plaster to rise above the edge of the shell.

3. Let the experiment set overnight. The next day, gently pry the shells loose. You will see shell depressions in the plaster. Spread a thin coat of petroleum jelly into these areas.

4. In another margarine container, mix up a new batch of plaster. Pour the fresh plaster into the depressions and let it harden overnight. The next day, lift the small plaster pieces from the surface. They will be exact models of the seashells.

Explanation:

You have just performed the process by which a *fossil* – an impression of a plant or animal of the past that has been preserved in the earth's crust has been created. A dying plant or animal is covered with mud, which hardens around its shape. As the matter decays and the cavity fills with minerals, a copy of the original plant or animal is formed from the minerals. Your experiment took few days, but real fossils are formed over hundreds of years. Scientists had been studying fossils from long ago to learn what kind of plants and animals were alive during the Earth's long history. The study of fossils is called *paleontology*.

61 An Ant Colony

Things Required:

- 2 glass plates
- 4 wood strips
- A plastic tape
- Cotton
- Shovel
- White cloth
- Sponge
- Sugar water
- Honey
- Black construction paper
- THE HELP OF YOUR PARENTS OR TEACHER

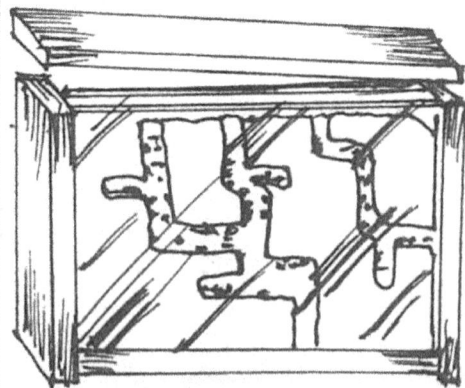

Directions:

This experiment will be great fun during your summer vacation! Build an ant farm, with the help of an adult, from 2 plates of glass and wood strips. The exact measurements will depend upon the materials you have on hand, but the sheets of glass should be placed about 2.5 cm apart.

1. With the glass plates 2.5 cm apart, tape the wood strips around the glass plates, making a rectangular glass and wood container. Do not tape the top wood strip to the glass plates yet.

2. Ask one of your parents do drill a hole in the top wood strip. This hole will be used to water the soil - the soil must be kept moist at all times. Plug the hole with some cotton between waterings.

3. Dig up an ant hill, including the area around it. Spread the soil on a white cloth to locate the queen. She will be much bigger than the other ants. Place the soil, queen, and other ants inside their new home.

4. Place a wet sponge, a tiny, open container of sugar water, and a few drops of honey on top of the soil. Then tape the remaining strip of wood - the one with the hole in it - to the top of your ant farm.

5. If you cover the glass with black construction paper for the first two weeks, it will encourage the ants to burrow next to the glass, and you will be able to see their tunnels clearly. You will soon discover many interesting facts about your ants.

Explanation:

You are observing an *ant colony*. Ants live and work in a society - an organised group of individuals, in which all the members follow rules. An ant society has one queen and many worker ants, and they all have special jobs. Study your ant farm to see the many different jobs the ants perform!

62 Cracker in the Mouth

Things Required:

✎ Unsweetened cracker

Directions:

1. Place an unsweetened cracker in your mouth. Chew it thoroughly, but don't swallow it.

2. Continue to chew without swallowing for several minutes. Does the cracker now taste sweet?

CHOMP!

CHOMP!

CHOMP!

CHOMP!

Explanation:

The moisture in your mouth is called *saliva*. Saliva contains chemicals that start to break down the food before it enters your stomach. The cracker has been changed by the action of these chemicals into simple sugars that your body can use for energy, and you can taste this sugar while the cracker is in your mouth.

63 A Muscular Squeeze

Things Required:

- 2 cherry cough drops
- 2 small jars, such as baby food jars, with lids
- Water

Directions:

1. Place a cough drop in each small jar, and then fill each about halfway full with water.
2. Screw the lids on tightly. Gently shake 1 jar by turning it upside down and then right-side up, over and over again. Leave the other jar alone.

3. After several minutes, notice the colour of the water in the 2 jars. The jar that you shook contains water of a deeper colour than the jar that remained quiet.

Explanation:

When the jar is shaken, the motion of the water helps to dissolve the cough drop more quickly. When you eat, your stomach acts the same way. It does not sit still, but rather the muscles squeeze and churn the food so that it breaks apart into pieces and becomes watery.

64 From Top to Bottom

Things Required:

- Yourself
- Asparagus
- Bathroom

Directions:

1. The next time you have asparagus for dinner, note its distinctive odour and taste, and eat a generous helping.

2. When you go to the bathroom the next morning, do you notice the same asparagus smell? Why does this happen when other foods that you eat do not produce their own odour in your urine?

Explanation:

A substance's odour is produced by molecules that your nose detects. The molecules that produce the smell of asparagus enter your body when you eat the vegetable, and are absorbed through your small intestine. However, your body does not use these particular molecules and they are passed, unchanged, into your kidneys. They finally leave your body in your urine.

65 Means of Identification

Things Required:

- A soft-lead pencil, such as a No. 2 pencil
- 2 sheets of paper
- Your finger
- Transparent tape

Directions:

1. Hold the pencil on its side and rub the lead on a sheet of paper. Continue rubbing the same spot over and over until it is covered with a heavy black coating.

2. Now, press your index finger into the lead spot and move it around. Your fingertip should become thoroughly smeared with pencil lead.

3. Press your blackened fingertip onto the sticky side of a small piece of transparent tape. Slowly peel away the tape and stick it onto a clean sheet of paper. You will see an interesting pattern of lines and swirls.

Explanation:

You have just taken your *fingerprint*. The pencil lead is transferred to your fingertip when you rub your finger in it. Then the sticky tape picks up the dark colour from your finger's ridges. No one else has the same fingerprints as you! Long ago, people would sign their letters by placing a thumbprint on the paper, and about fifty years ago in the United States, people started using *fingerprints as a means of identification*. And once, a man surveying a land in New Mexico used his thumbprint on his reports so that others could not forge his name!

66

Head First

Things Required:

- A friend
- A chair

Directions:

1. Ask a friend to relax in a chair. He should fold his arms and stretch his feet as far forward as possible. Tell him to lean far back so that his head faces upward.

2. Now press a fingertip onto his forehead and challenge him to get up from the chair without unfolding his arms or moving his feet. He will struggle, but will not be able to rise.

I CAN'T MOVE!

Explanation:

Your friend must gain his balance before getting up from the chair. To do this, he must raise his head first. However, you are pressing down on his forehead and this prevents him from taking the first action.

67 Unequal Pressure

Things Required:

✐ Yourself

Directions

1. Pinch your nose closed with your fingers.
2. Now swallow. Your ears will feel blocked. Have you ever had this feeling before?
3. Take a big yawn. Your ears will return to normal.

CAN'T HEAR!

YAM!

Explanation:

There is a stretchy tissue inside each of your ears that helps you to hear. This tissue is called an *eardrum*. Sometimes, there is a difference in air pressure outside and inside your head, such as when you climb a mountain or go up in an elevator. To protect your eardrums from breaking under this kind of increase in pressure, a tube – *the eustachian tube* – that leads from each ear to your nose and throat allows air of the same pressure as the outside to get behind your eardrums. When you blocked your nose and swallowed, you actually blocked your eustachian tubes and created an unequal pressure in your head. By yawning, you returned everything back to normal.

68 Look into My Eyes

Things Required:

- ✏ 10 sheets of typing paper
- ✏ Rubber band
- ✏ This book or any book
- ✏ Mirror

Directions:

1. Roll the 10 sheets of typing paper into a hollow tube. Slip a rubber band over the roll to hold it in place.

2. Hold the tube against some words on this page and look into the cylinder. You face should be pressed against the tube so that no light enters. At first, you will not be able to see much. However, your eye will soon adjust to the dark and allow you to read the words.

3. Lift your head and immediately look at your eye in the mirror. Do you see any changes?

Explanation:

The part of your eye into which light enters is called the *pupil* - the dark spot in the centre of your eye. The pupil gets bigger in the dark because it must allow more light in to help you see better. When the light is too bright, the pupil tries to shut some of it out by becoming smaller.

69 A Total Picture

Things Required:

- A paper towel tube
- Matchbook

Directions

To discover a power that you are probably not aware of, try this easy experiment. Did you know that your eyes can pierce solid objects?

1. Hold an empty paper towel tube in front of one eye. Hold a matchbook in front of the other eye, with the cardboard touching both the tip of your nose and the tube.

2. Look straight ahead while keeping both of your eyes open. Your eyes have just burn a hole through the matchbook!

Explanation:

You use both your eyes to see an object. *Your brain receives a message from each eye and combines them for the total picture.* Usually the messages are the same because both eyes are looking at the same thing. In this experiment, however, each eye sees a different image: One eye sees the matchbook and the other sees the view beyond it. The brain combines these and it seems as if you are looking right through the matchbook.

70

A Cold Hand

Things Required:

- 3 bowls
- Water

Directions:

1. Fill 3 bowls with tap water as follows: One with cold water, one with lukewarm water, and one with hot water – but not hot enough to burn. Place the bowls on a table or a counter, with the lukewarm water between the other two.

2. Place one hand in the cold water and one in the hot. Let your hands adjust to

the temperature for several minutes. Then take them out and plunge them both into the bowl of lukewarm water. The hand that was resting in the cold water now feels warm and the other hand feels cold. Do you know why?

Explanation:

The hand that was resting in the cold water is now placed in warmer water and some heat from the water is transferred to the skin. Therefore, the hand feels warm. The opposite happens with the other hand - it is now warmer than the surrounding water and some heat moves into the water. As a result, the hand feels cold.

71 Seaweed Collection

Things Required:

- Pail
- Seaweed
- Shallow pan
- Water
- Finger-paint paper (paper with one glossy side)
- Glass pane, smaller than the shallow pan
- Scissors
- Pencil
- Paper toweling
- Book

Directions:

If you live near the ocean or if you go down to the sea during the summer or on a vacation, you can make an interesting picture from seaweed.

1. Take a pail to the beach and collect some seaweed. Look for various types of green, brown, and red seaweed.

2. When you get home, fill a shallow pan about halfway with water, then cut the finger-paint paper to the same size as the glass pane.

3. Set the glass pane in the pan of water and press the paper, glossy side up, onto the glass. Make sure at least 6 mm of water covers the paper.

4. Now, arrange your seaweed on the paper. The layer of water will support the plants and allow them to assume their natural shape. Use the tip of a pencil to position them wherever you want on the paper.

5. Spread some paper towels on a flat surface. Slowly and gently lift the glass pane from the pan. (Do not tilt it or the seaweed will run off the edges.) Carefully slide the paper from the glass onto the paper toweling.

6. Place 2 or 3 paper towels over the seaweed and lay a heavy book on top. Let the experiment sit overnight. The next day, when you take away the book and paper towels, you will find a beautiful picture.

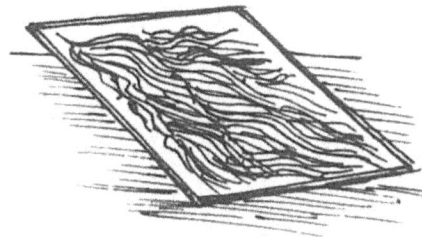

Explanation

Seaweed is a collection of living plants that grow in the ocean. The seaweed contains a type of natural glue that helps it to stick to the white paper. After you arranged the seaweed on the paper and the paper towels absorbed the excess water, the plants remained firmly attached to the paper.

10 Projects in Colour

1 Dry Matchsticks

Things Required:

- Matches or Matchsticks
- Newspaper
- Tall drinking glass
- Pail
- Water
- THE HELP OF YOUR PARENTS/GUARDIAN/TEACHER

Directions

Can you dunk some matches in water and still be able to use them?

1. Wrap a few matches or matchsitcks in a small scrap of newspaper. Crumple the newspaper and poke it to the bottom of a tall drinking glass. The paper should remain in the bottom when the glass is turned upside down.

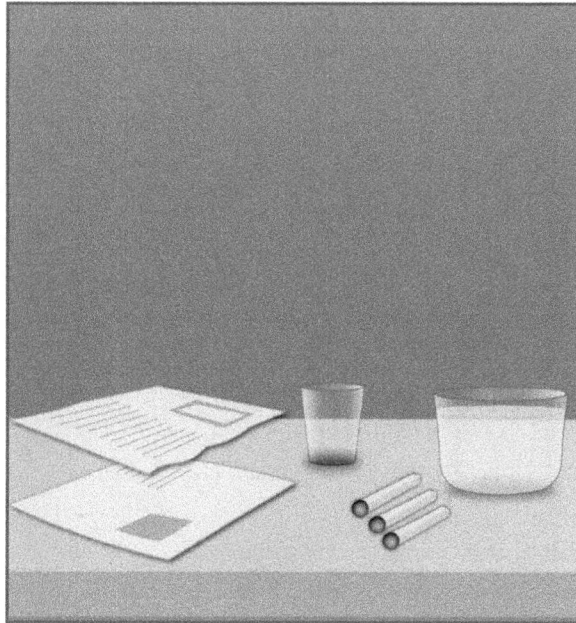

2. Fill a pail, or other deep container, with water. Hold the glass upside down and push it straight down to the bottom of the pail. Make sure you don't tip the glass sideways.

3. Now, remove the glass, take out the newspaper, and unwrap the matches. The newspaper and the matches are dry! Prove it by asking one of your parents to strike the matches.

Explanation:

When you pushed the glass into the water, the glass was not really empty. It was full of air. This volume of air prevented water from entering the glass, and the paper and matches remained dry.

2 Dishwashing Detergent

Things Required:

- Light-coloured bowl
- Water
- Paprika
- Dishwashing liquid

Directions

1. Fill a light-coloured bowl with water and shake some paprika evenly over the top of the water.
2. Put a drop of dishwashing liquid on your finger, and then dip your finger into the centre of the bowl. The red paprika quickly scoots to the sides of the bowl.

Explanation:

Dishwashing liquid is a detergent, and one of the important qualities of a detergent is that it mixes easily with water. As you dipped your finger into the bowl, a

small amount of dishwashing liquid from your finger readily attached itself to the water. Then it quickly spread over the entire surface and pushed all the grains of paprika to the sides of the bowl.

3 A Forward Thrust

Things Required:

- 2 antacid tablets
- An empty plastic bottle, such as a 15 cm-tall shampoo bottle
- Water
- A pan

Directions

1. Break apart two antacid tablets and put the pieces in an empty plastic bottle.
2. Now fill the bottle about full with water. Rest the bottle on its side in a pan of water. You will see the bottle-boat chug along the water's surface.

Explanation:

When the antacid tablets and water combine, they form a gas that escapes through the neck of the bottle. This backward motion of the escaping gas is matched by an equal forward thrust, which propels the boat ahead.

4 Hidden Power

Things Required:

- A friend
- Broom

Directions:

Challenge your friend with this experiment - you'll always win!

1. Ask your pal to hold both hands straight out and grasp the broom.
2. Place one of your hands in the centre of the broom, with your arm bent at the elbow, grasping it with a slight downward tug. Tell your friend that you bet he or she can't push you over with the broom.

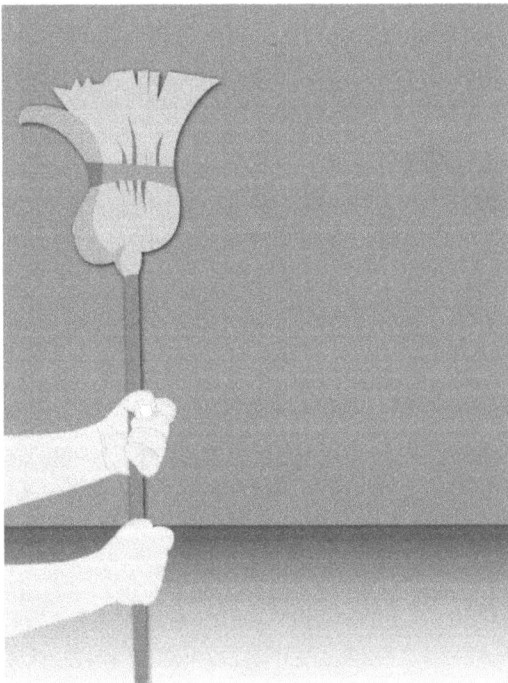

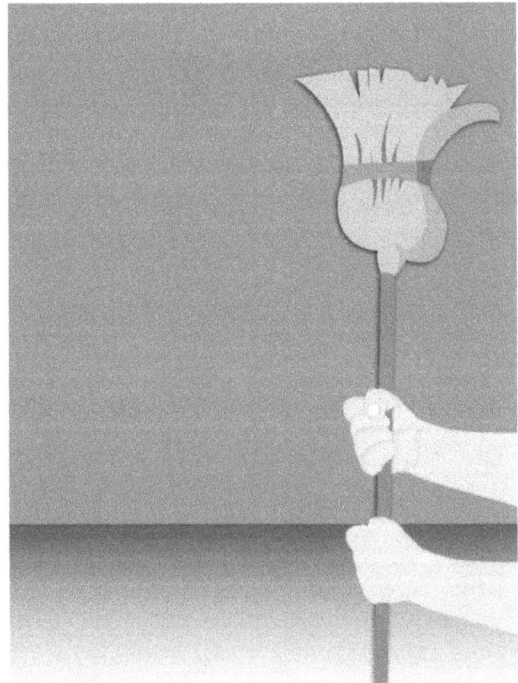

3. As your friend pushes the broom towards you, push straight up. You will remain standing in your place fixed without movement.

Explanation:

Even if your friend is bigger and stronger than you, you will always win because you have a hidden power. You have much more *leverage - with one bent arm acting as a lever -* than your friend has with two straight arms. A lever helps to lift weights with less effort, giving you a mechanical advantage. So the direction of your friend's pushing force is easily overpowered by a much smaller force from you.

5 Centre of Rotation

Things Required:

- 4 small plastic bowls
- A strong plastic tape
- 16 marbles
- A meter long board
- Several books

Directions

1. Tape 2 small plastic bowls together back to back, forming a wheel. Make another wheel with 2 more identical bowls.

2. Tape 4 marbles to the centre of each side of one wheel. Then tape 4 marbles to each side of the other wheel, but place 2 marbles close to the inside rim and 2 marbles directly opposite.

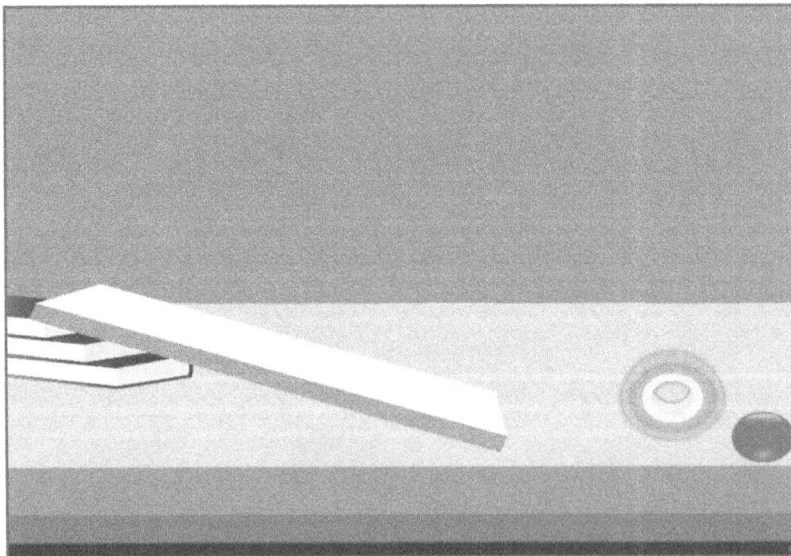

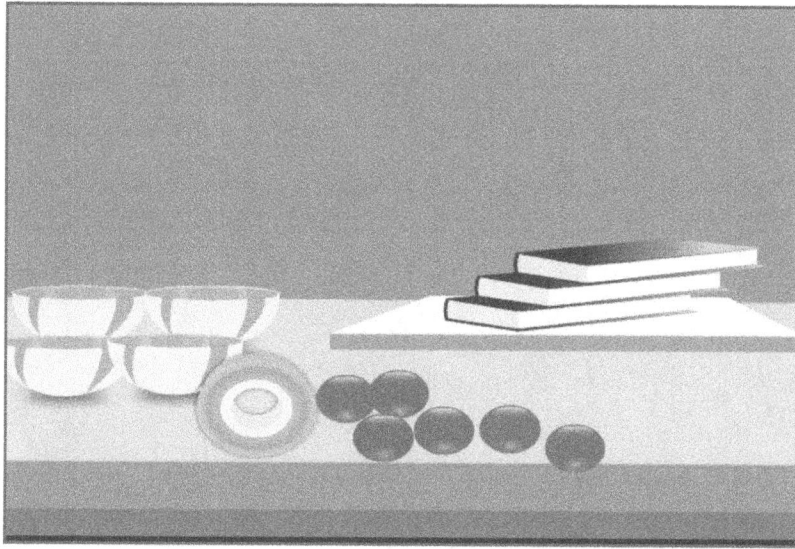

3. Prop up the end of a board with several books. Hold both the wheels at the high end.

4. Release the wheels at the same instant. The wheel with the weights in the centre will speed ahead of its opponent. Can you explain why?

Explanation:

The wheel with the marbles placed at the rim loses speed because energy must be used to move the weight of the marbles. The faster wheel, however, has the weight of the marbles at the exact centre of rotation and does not waste energy spinning the marbles around the outer edge.

6 Hydroelectric Power

Things Required:

- A pencil
- A paper plate with ridges along the edges
- Water faucet

Directions

1. Push a pencil through the centre of the plate, wriggling the pencil back and forth to make the hole loose.

2. Turn on the tap water to produce a steady stream. Hold the pencil so that one edge of the plate touches the water. The plate will spin. Turn the faucet higher and the plate will spin faster.

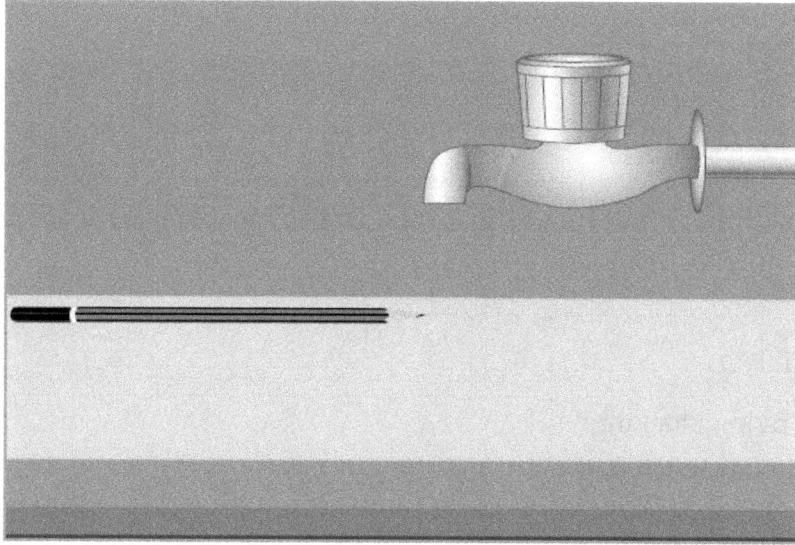

Explanation:

You have just built a *waterwheel*. Waterwheels many times the size of your model are used in large rivers and near waterfalls. The water causes the waterwheel to turn, and the turning motion generates electricity. We call the energy that is made this way as *hydroelectric power*.

Repelling Charges

Things Required:

- An old nylon stocking
- A plastic sandwich bag

Directions:

1. With one hand, hold the toe of a nylon stocking (the shiny kind works best) on the edge of a table. Slip your other hand into the plastic sandwich bag and grip the stocking, starting at the toe and pulling along the entire length, up to the thigh. Repeat this process several times, stroking the nylon in the same direction.

2. Holding the top edge of the stocking, lift it from the table. Let go with the bagged hand. Suddenly the stocking will appear as if there were a leg in it.

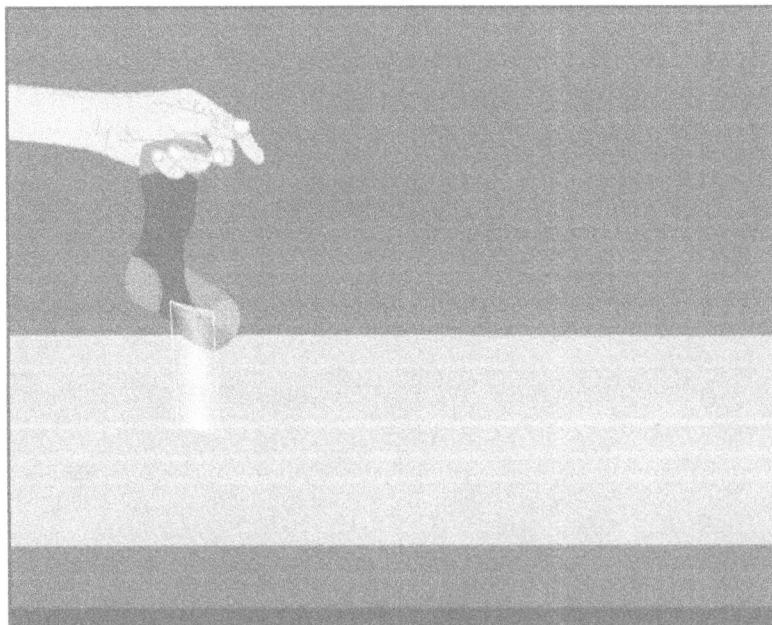

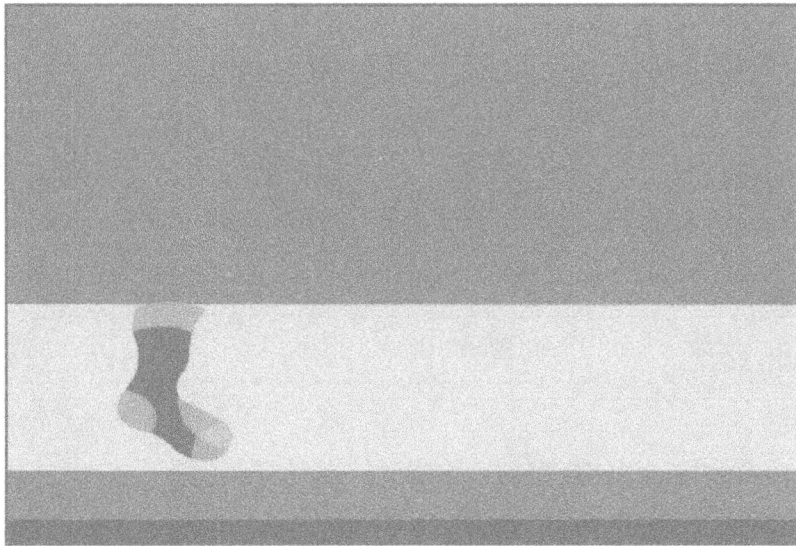

Explanation:

When you stroked the nylon with the plastic bag, these two materials acquired *opposite electrical charges*. Because the stocking material *became all the same charge and because like charges repel each other*, the sides of the stocking pushed away from each other, taking on the shape of a real leg.

8 An Invisible Gas

Things Required:

- Candle
- An aluminium pie plate
- Matches or Matchsticks
- 1 teaspoon baking soda
- Glass (not plastic) measuring cup, marked for millilitres
- Vinegar
- THE HELP OF YOUR PARENTS, GUARDIAN OR TEACHER

Directions

1. Set a candle on an aluminium pie plate in the sink. Then ask one of your parents to light the wick.

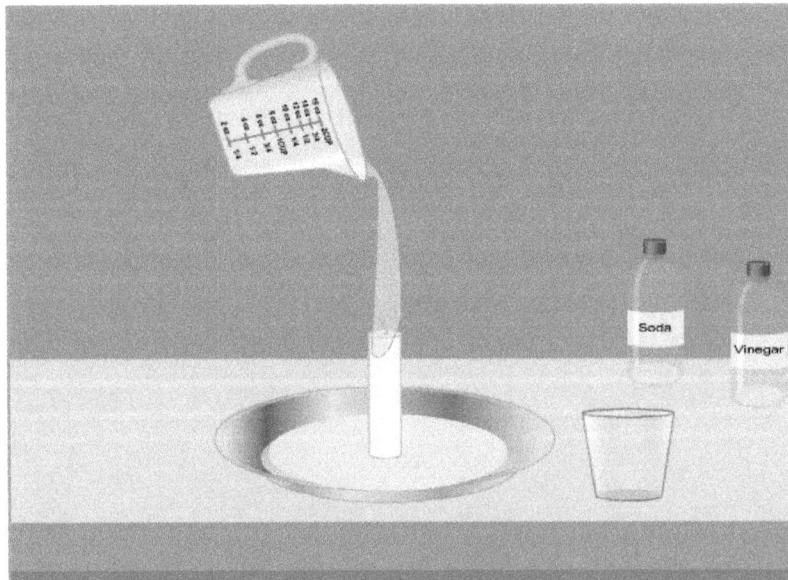

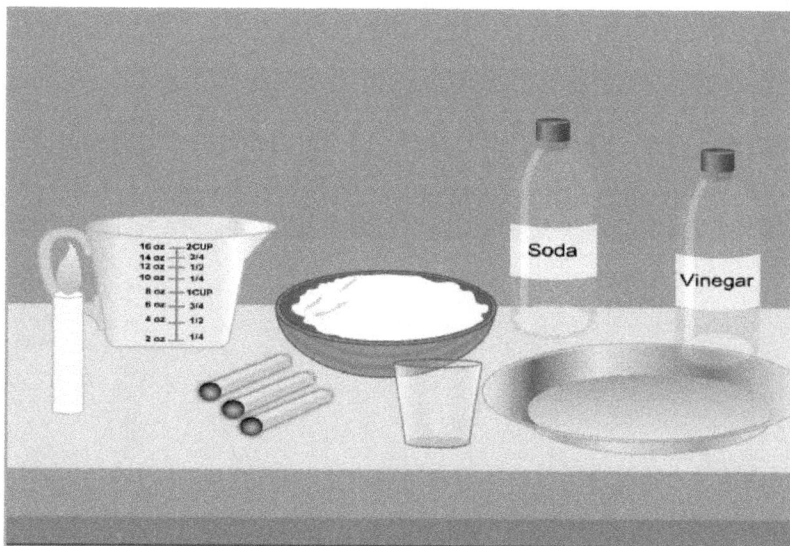

2. While the candle is burning, place the teaspoon of baking soda into a spouted measuring cup. Look at the markings and notice where the 30 ml line is drawn. Pour vinegar to this mark. The mixture will foam up for a few seconds.

3. As soon as the bubbling has settled down a bit, slowly lift the measuring cup and hold the spout a few centimeters directly over the flame. Tilt the cup forward as if your were going to pour out the contents, but do not let any solution dribble out. The flame will go out.

Explanation:

The baking soda and vinegar reacted to form *carbon dioxide gas* (CO_2). This gas is invisible. It remained in the measuring cup until you tilted the spout forward. Then it flowed from the cup onto the flame because carbon dioxide is heavier than air. The flame was smothered with carbon dioxide and went out.

9 A Spider's Web

Things Required:

- A spider web
- White paper
- Safety pins

Directions

1. Find an old spider web with some dead insects left in the threads – you might look in a garage, on a porch, or on some trees or bushes.
2. Pull one of the dead insects from the web and place it on a sheet of white paper.
3. With a safety pin or other small, pointed instrument, break apart the insect's

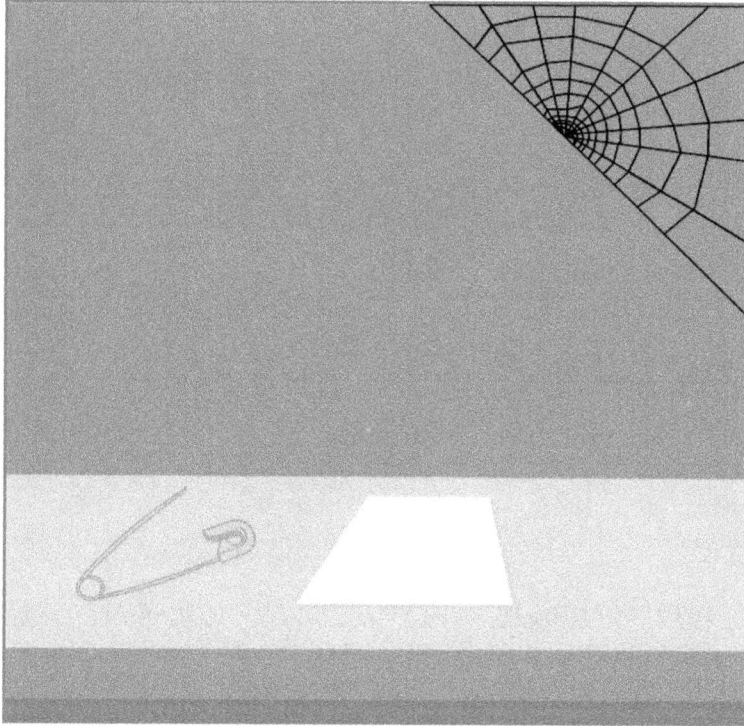

hard outer shell. You will see that there is nothing left inside. How did the spider eat the inside of his catch without opening the shell?

Explanation:

When an insect gets caught in a spider's web, the spider spines extra thread around its victim to hold it in place, but he does not have to do any chewing through the tough shell. The spider punctures the insect with sharp fangs and injects a chemical. This chemical makes the insect's insides soft and watery, and then the spider sucks up its meal.

10 Beats per Minute

Things Required:

- A wooden match
- A thumbtack, with a wide, flat surface

Directions:

You can feel your heart beating, but did you know that you can also see it beating? All it takes is a simple device, which you can make in a few minutes.

1. Press a wooden match onto the point of the thumbtack.

2. Set your device on top of your wrist and move it from place to place until you find a strong beat - your pulse. Your device will respond by tick-tocking back and forth like a grandfather clock pendulum.

3. Count how many times the wooden match moves in one minute.

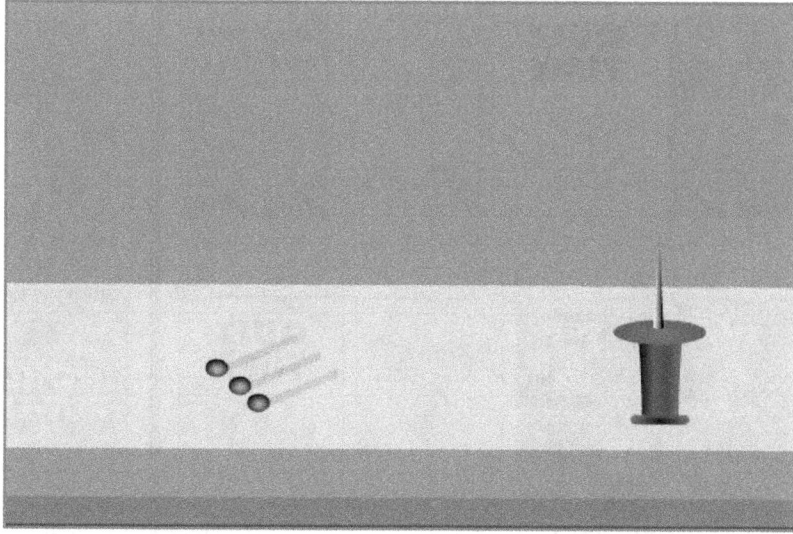

Explanation:

When you measure how many times your heart beats, you are taking your *pulse*. Blood is pumped by your heart throughout your entire body, carried by arteries and veins. Some veins are close to the surface in your wrist, and this is a good place to measure the beating action. You probably obtained a count between 90 and 120 beats. *As you grow older, your heart will slow down to about 80 beats per minute.*

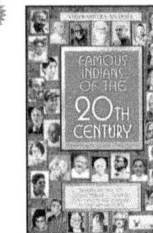

STUDENT DEVELOPMENT/LEARNING

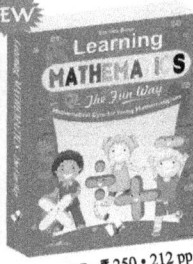

02502 P • ₹ 100 • 112 pp

02206 P • ₹ 100 • 128 pp

03402 P • ₹ 195 • 208 pp

03401 P • ₹ 250 • 212 pp

00503 P • ₹ 135 • 142 pp   10501 P • ₹ 96 • 152 pp   00507 P • ₹ 150 • 133 pp

9076 D • ₹ 80 • 144 pp   10502 P • ₹ 96 • 144 pp

PUZZLES

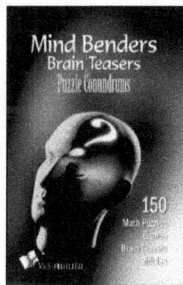

02311 P • ₹ 96 • 112 pp

12302 P • ₹ 48 • 112 pp   02305 P • ₹ 60
96 pp

02306 P • ₹ 60
96 pp

02301 P • ₹ 110 • 152 pp

DRAWING BOOKS

12501 P • ₹ 150
122 pp (with CD)

02501 P • ₹ 150
128 pp (with CD)

02503 P • ₹ 295 • 108 pp

12506 P • ₹ 120 • 84 pp   025051 P • ₹ 120 • 84 pp

POPULAR SCIENCE

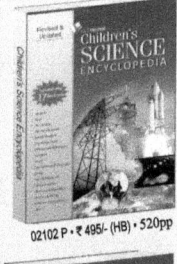

12141 P • ₹ 495/- (HB) • 520pp   02102 P • ₹ 495/- (HB) • 520pp

12103 P • ₹ 120 • 148 pp   02103 P • ₹ 120 • 148 pp

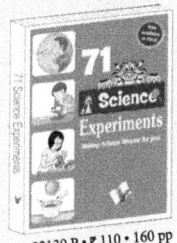

12140 P • ₹ 110 • 160 pp   02139 P • ₹ 110 • 160 pp

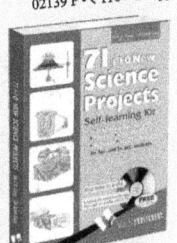

12101 S • ₹ 160 • 136 pp
(Available in Tamil, Bangla)

02101 P • ₹ 160 • 120 pp

02212 P • ₹ 100 • 124 pp   02201 P • ₹ 80 • 44 pp

VALUE PACKS

(12410S)   (00608S)   (02312S)   (00223S)   (14001S)   (10505S)   (12211S)

Contact us at sales@vspublishers.com